DR. PIERRE F. WALTER

BASICS OF FENG SHUI

A Beginner's Guide

"Articles Series"

©2011 Pierre F. Walter. All rights reserved.

Published by Sirius-C Media Galaxy LLC

http://sirius-c-publishing.com

http://siriuscmedia.com

http://ipublica.com

ISBN 978-1-468125-94-8

Contact Information Dr. Pierre F. Walter

publisher@sirius-c-publishing.com

About Dr. Pierre F. Walter

http://drpfw.info

Quotation Suggestion

Pierre F. Walter, *Basics of Feng Shui: A Beginner's Guide,* Newark: Sirius-C Media Galaxy LLC, 2011

About the Author

Pierre F. Walter is an author, international lawyer, researcher, corporate trainer, and lecturer. After finalizing studies in German Law, International Law and *European integration* with diplomas obtained in 1981 through 1983, he graduated in December 1987 at the Law Faculty of the University of Geneva as *Docteur en Droit* in international law.

The doctorate was funded by scholarships from the *Swiss Institute of Comparative Law*, Lausanne, and from the *University of Geneva*, as well as a Fulbright Travel Grant for an assistantship with Professor Louis B. Sohn at *UGA Law School Department of International Law*, Athens, Georgia, USA, in 1985. Pierre F. Walter also served as a research assistant to *Freshfields, Bruckhaus, Deringer*, Cologne, Germany in 1983 and to *Lalive Lawyers*, Geneva, in 1987.

Pierre F. Walter writes and lectures in English, German and French languages; he has written *more than ten thousand pages* embracing all literary genres, including *novels, short stories, film scripts, essays, selfhelp books, monographs* and extended *book reviews*. Also a pianist and composer, he has realized 40 CDs with *jazz, newage* and *relaxation music*.

Pierre F. Walter's professional publications span the domains *International Law, Criminal Law, Holistic Science, Psychology, Education, Shamanism, Ecology, Spirituality, Quantum Physics, Systems Theory, Natural Healing, Peace Research, Personal Growth, Selfhelp* and *Consciousness Research*. 110 Book Reviews, thirty-eight audio books and more than hundred video lectures were realized in the years 2005-2010. Besides, Pierre F. Walter is author and editor of *Great Minds Series*, which features scientists, artists and authors of genius from Leonardo to Fritjof Capra.

Pierre F. Walter publishes via his Delaware firm *Sirius-C Media Galaxy LLC* and the imprints IPUBLICA and Sirius-C Media (SCM).

For Nelson

CONTENTS

SOME BASICS   7
What is Feng Shui?

    What is Feng Shui?   8

    Yin and Yang   11

    The Five Elements   13

FENG SHUI BOOK REVIEWS   14

    7 Feng Shui Books Reviewed

    Introduction   15

    The Book Reviews   19

Amazing Scientific Basis of Feng Shui

The Design of Feng Shui Logos, Trademarks & Signboards

Creating Sacred Space with Feng Shui

Feng Shui

The Feng Shui Kit

    First ring (I Ching sign): Chen

    Second Ring (Direction): East

    Third Ring (Element): Earth

Fourth Ring (Yin/Yang): Yang

Fifth Ring (Animal): Dragon

The Feng: The Book of Cures

The Feng: Feng Shui Book and Card Pack

BIBLIOGRAPHY 49
Contextual Bibliography

FROM THE SAME AUTHOR 81
A Bibliography

SOME BASICS

What is Feng Shui?

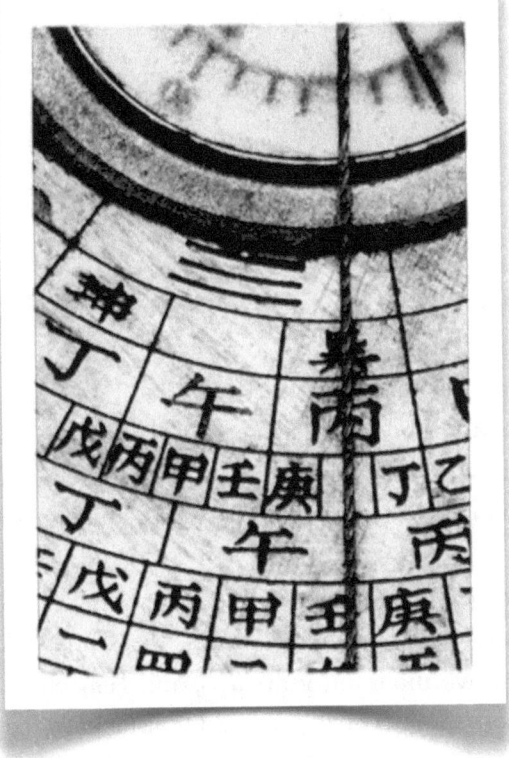

What is Feng Shui?

Feng Shui is *not* a belief system, as astronomy is not a belief system. Feng Shui is a science of the bioenergy and perhaps the oldest distillation of this *holistic knowledge* into something we today would qualify as a science while Asians rather speak of philosophy or of religion when they talk about the perennial science of the *bioenergy*.

While Feng Shui is practiced as a business in Hong Kong, Kuala Lumpur and Singapore, for construction of houses and offices, it has to be seen that this is *only the pragmatic branch* as it were of a larger body of science that really deals with all living and how human beings relate to the greater realms of plants, animals, water, rivers, mountains, and generally to the cosmos.

While *systems theory* in the West only now and gradually reveals our total interconnectedness, the Chinese knew this more than five thousand years ago, and Feng Shui is one of many manifestations of this age-old holistic knowledge.

However, even in the West, alternative scientists have acknowledged the existence of the bioenergetic functionality not only of the human organism, but also of the weather, the atmosphere and the cosmos as a whole. While in substance all these researchers observed basically the same phenomena, the way they named the bioenergy varied. Paracelsus, for example, spoke of *vis vitalis*, Swedenborg of *spirit energy*, Mesmer of *animal magnetism*, Reichenbach of the *odic force*, Lakhovsky of *universion* and Wilhelm Reich of *orgone*.[1]

And since millennia this same energy was called *ch'i* by the Chinese, *ki* or *hado* by the Japanese, *prana* in ancient India and *mana* with the Kahunas from Hawaii and the Tsalagi natives of North America.[2]

Finally, parapsychologists and aura healers universally agree that the motor of all psychic phenomena is to be found in our bioplasmatic and egg-shaped aura, the *luminous energy body* of low density that we carry around our physical body and which could be seen as an extension of our bioplasmatic energy because it is composed of the same bioenergetic charge that we can directly detect within the bioplasma.[3]

Emotions, as I have shown in my audio book *Emonics (2010)* as well as my *Idiot Guide to Emotions (2010)*, are energy streamings that are directly related to the bioenergy and thus

---

[1] See Pierre F. Walter, *The Science of Orgonomy, Monograph (2010)* and *Energy Science and Vibrational Healing, Monograph (2010)* as well as my audio books *Emonics (2010)* and *Emotional Flow (2010)*.

[2] See, for example, Max Freedom Long, *The Secret Science at Work: The Huna Method as a Way of Life (1953/1995)* and Erika Nau, *Self-Awareness Through Huna (1981)*. Regarding the notion of *hado*, see Masaru Emoto, *The Hidden Messages in Water (2004)* and *The Secret Life of Water (2005)*.

[3] See Dean Radin, *The Conscious Universe (1997)* and *Entangled Minds (2006)*, as well as Shafica Karagulla, *The Chakras (1989)* and Charles W. Leadbeater, *The Inner Life (1911)*.

have their place not in the brain, as so-called cognitive elements, but in the bioplasma and in the aura.

I got in touch with Feng Shui in 1996, while living and working in Asia. I have researched quite a bit, since then, about this *perennial science* and have written book reviews of major Feng Shui books and card sets, which I will reprint in this volume.

Basic to Feng Shui is the insight that only harmony can ensure that nature keeps a balance between growth and stagnation, as unrestrained growth is cancerous and insufficient growth leads to stagnation. The secret is a dialectics between charge and discharge, male and female, or, generally put, *yin and yang* as the magnetic poles of the total sphere of the living process.

Yin and Yang

The science of Feng Shui can only be understood by the Western reader once he or she is familiar with the energy duality of *yin* and *yang*, which are manifestations of *energy polarity* within the cyclic movements of the human energy field.

The concepts of *yin and yang* originate in ancient Chinese philosophy and metaphysics, which describes two primal opposing but complementary forces found in all things in the universe. Yin is the darker element; it is sad, passive, dark, feminine, downward-seeking, and corresponds to the night, flat land, the moon, shadow and receptivity. Yang is the brighter element; it is happy, active, light, masculine, upward-seeking and corresponds to the day, the sun, mountainous landscapes, light and creativity. Yin is often symbolized by water or earth, while yang is symbolized by fire, or wind. Yin and yang are descriptions of complementary opposites rather than absolutes.

Any yin-yang dichotomy can be seen as its opposite when viewed from another perspective. Most forces in nature can be seen as having yin and yang states, and the two are usually in movement rather than held in absolute stasis. In Western culture, yin and yang are often inaccurately portrayed as corresponding to evil and good or reduced to a simplistic reductionism male-female. Yin can well be associated with the female principle but this does not mean that it is identical with it. It's actually a bit like in the cabalistic system. We talk about *corresponding characteristics* or elements, and the system as such is one of corresponding relationships.

Yin can be said to correspond with water, the female principle, the color black, the direction down or a landscape that is flat. Yang can be said to correspond with fire, the male principle, the color white, the direction up or with a landscape that is mountainous.

In every yin there is a bit of yang, and in every yang a bit of yin. This bit is the essence that is multiplied once the point of culmination has been passed. What that means is that for example yin moves towards its fullness in order to culminate and swap its nature into yang.

Yang, when it culminates, becomes yin. That is why we can say change is programmed into the very essence of the *yin-yang dualism* and thus, change cannot be avoided. We can even go as far as saying that the very fact of change is the proof that we deal with a living thing. If there is no change, there is no movement and, as a result, no life. Life is change, living movement.

The Five Elements

The principle of the five elements, which is an integral part of the science of Feng Shui, suggests that nature is interactive and in a continuous process of transformation. The five elements wood, fire, water, earth and metal are *mutually constructive* and also *mutually destructive*. For example, wood is positively enhanced by water whereas water destroys fire. These two parallel processes of creation and destruction can be seen as two circles or cycles, a cycle of creation, and a cycle of destruction.

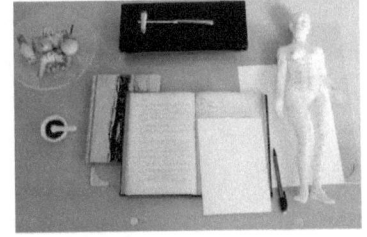

The principle of the five elements teaches us that nothing in nature is static, but that all is subject to continuous flow, continuous change. It also teaches us that all elements naturally interact with each other, as they mutually depend on each other, and that nothing is really isolated.

As a result, by studying and observing these laws, we notice a high degree of interdependence and high interactivity, a fact that in Western science has only recently been given the focus it merits. It is modern *systems theory* that deals with the interactive processes in nature.

The principle of the five elements, as simplistic as it may seem on first sight, is a wonderful lesson about the functional worldview of nature, and as such it is the point of departure of a holistic view of life.

FENG SHUI BOOK REVIEWS

7 Feng Shui Books Reviewed

Introduction

Dr. Ong Hean-Tatt, a bioenergy researcher from Malaysia, wrote a concise study about the scientific basis of Feng Shui, the five thousand years old energy science of the Chinese and concluded from a wealth of observations and discoveries that this science deals with the cosmic energy using about the same precision and objectivity as Newtonian physics regarding gravity.

In addition, as I have shown in my review of this important book, Dr. Ong establishes amazing parallels between Feng Shui and the *perennial knowledge* about the telluric  force known as *geomancy*, which has a long-standing tradition in both the East and the West.

The factual evidence produced by the author that relates in detail to various UFO sightings and reports from reputed sources is dumbfounding and seems to prove the fact that these phenomena feed upon earth energies or telluric energies emanating from underground water. He also found that important religious cult sites, such as Stonehenge, were built exactly on the intersection of telluric lines. And not astonishingly so, it's around these sites that most of spirit, angels, ghost and UFO sightings actually occur, and for the very reason that these places are flooded with cosmic energy and therefore allow other dimensions to connect with ours through energetic cross-section and vibrational resonance.

Further, Dr. Ong examines the bird migration phenomenon and finds that it corroborates the evidence forwarded for the existence of the telluric world grid – the fact is that the birds more or less follow those lines and that the energy that emanates from them serves the birds as a navigation help.

In his conversations with Bill Moyers, Joseph Campbell speculates that all gods in all religions are ultimately but energy manifestations:

Joseph Campbell

[T]he gods are rather manifestations and purveyors of an energy that is finally impersonal. They are not its source. The god is the vehicle of its energy. And the force or quality of the energy that is involved or represented determines the character and function of the god. There are gods of violence, there are gods of compassion, there are gods that unite the two worlds of the unseen and the seen, and there are gods that are simply the protectors of kings or nations in their war campaigns. These are all personifications of the energies in play. But the ultimate source of the energies remains a mystery.[4]

---

[4] Joseph Campbell, *The Power of Myth (1988)*, p. 259.

Feng Shui, now booming as the first and foremost technique of living in harmony with cosmic energies, is by its very nature alien to Western thought. This is so because in the West, the perception of life as an energetic process that is highly complex yet also highly flexible became taboo with the dominance of powers that tried to control life and man.

Knowledge about life was forbidden and left to alchemists who risked their lives pursuing the only real science that existed at that time. Even with the so-called scientific era of humanity, the knowledge taboo persisted and the dominance only changed its camps. What formerly was the privilege of Church officials, was then, and is until today, in the hands of scientists that are funded by the military and huge multinational corporations. Hence, the saddening result that official Western science is always ultimately system-conform.

In Asia, however, this knowledge-prohibition never existed and *Feng Shui*, the Chinese science of the cosmic life energy and its manipulation for health, power and happiness, is not only a set discipline for every scholar, but has its roots in popular tradition.

These book reviews show both the Eastern and the Western approach to the perennial science of *Feng Shui*, and it becomes obvious that this science is something that fascinates high-caliber scientists such as Dr. Ong from Kuala Lumpur as well as housewives who may pick a Feng Shui card to assess if the oven is best put in the corner or in the middle of the kitchen …

This, then, is again typical for Asian scientific concepts; they are never pure theory but have their practical stance in life, and if not, they won't be pursued at all. The problem is that Western thought is so much pervaded with the dichotomy science versus ordinary life that people in the West tend to judge these practical applications of Feng Shui in a depreciative way, saying that Feng Shui was 'the art of furniture arrangement' and other reductionist statements of this kind. When the science of Feng Shui explains how the vital energies move in our house, and accordingly advises to arrange furniture in a certain way, and not in a certain other way, then this is not a reason to deny this science its very status as a science, and relegate it to the areas design, lifestyle, fashion, favorite interests, and all the rest of the modern consumerist soup.

While it was admittedly designers who first discovered Feng Shui in the West, this doesn't mean that Feng Shui is *but* a design tool, as many people tend to affirm. That would be like saying that electromagnetics is not a science because a light bulb is 'after all a piece of furniture'.

But it is exactly this kind of reductionist thinking that has precluded the understanding of Asian scientific concepts for so long.

The Book Reviews

Dr. Ong Hean-Tatt, PhD
Amazing Scientific Basis of Feng Shui

Evelyn Lip
The Design and Feng Shui of Logos

Karen Kingston
Creating Sacred Space with Feng Shui

Lillian Too
Feng Shui

Man-Ho Kwok
The Feng Shui Kit

Nancilee Wydra
Feng Shui, The Book of Cures

Richard Craze

Feng Shui – Feng Shui Book and Card Pack

Dr. Ong Hean-Tatt

Amazing Scientific Basis of Feng Shui

Kuala Lumpur: Eastern Dragon Press, 1997
(No cover scan available)

Amazing Scientific Basis of Feng Shui is a treatise – nothing less. It is not your usual Feng Shui book. The analysis of Dr. Ong reveals that the science of Feng Shui is an integral part of a *perennial science concept* about the cosmic life energy that can be traced back, in Oriental cultures, until Antiquity. The strength of the book is the unusual and wide regard of the author on a matter that, in the trend of New Age enthusiasm, is sometimes treated in a rather esoteric and unscientific manner.

Dr. Ong's approach is deep, inter-disciplinary and synthetic. Furthermore, the book is rich in references and resources. The author must have done extensive research across various sciences to have reached at his amazing conclusions.

Before the Dr. Ong starts his cross-science survey, he carefully prepares the reader and introduces the concept of Feng Shui as a science about the life energy that can be compared to acupuncture and that possesses similar knowledge about the subtle energy that is at the basis of the now recognized medical treatment of acupuncture. Both concepts, Dr. Ong shows, are based upon the same point of departure: the existence of a polarized bioenergy that the Chinese call *ch'i*, the negative part of it being called *sha*.

Then, the author examines and explains the truth expressed in ancient myths and legends like Homer's Iliad and advances evidence that such stories are to be taken literal and not just as metaphorical, occult or esoteric.

The book proceeds logically, starting from the science of Feng Shui and its historical development in China, then extending to the broader concept of *geomancy* that Feng Shui is historically part of and that has also a long tradition in the West. Yet unlike typical Feng Shui books, the author does not enter the subject in practice, but focuses primarily on collecting and demonstrating scientific evidence, mostly from Western sources, in order to corroborate the view that Feng Shui is a *science* - and not superstition or magic, nor a design tool or fashionable lifestyle concept.

The regard that Dr. Ong takes upon the bioenergy is most interesting as he examines the traditional Chinese Ganzhi system, comparing it with the *Cabbalistic Sefirot* system, and then shows that Feng Shui can be compared with these sources of perennial wisdom and expresses the same truth in other words.

After an in-depth analysis of the I Ching, the *Pa Kua* (I Ching Based Feng Shui Compass) and the *Lo Pan* (Geomancy Based Feng Shui Compass), the author devotes three chapters to the examination of the concepts of ch'i, sha and the five elements - an invaluable source of knowledge especially for the Western reader and scientist. As a matter of fact, Dr. Ong shows the amazing similarity of the Chinese Science of the Five Elements with what he makes out as the Western Five Elements Concept. From there, he proceeds to examining the Western geomancy art of dowsing and provides evidence for the fact that there is a broad consensus between Chinese and Western sources regarding the detrimental effects of underground water and cancer-producing fault lines.

The factual evidence that Dr. Ong cites is of great value so much the more as these sources of evidence are difficult to make out.

But the author does not limit his research here. The way Dr. Ong explains the UFO phenomenon is among the most interesting and original approaches I have heard of until this day. Many of us may have doubted not the phenomenon as such, but the explanation for possible presence of extraterrestrial forces on earth. Actually, paranormals and scientists involved with parapsychology have since long questioned this theory and put forward their view that those appearances rather are concerned with earth-bound inhabitants of a different, generally higher, energetic vibration than humans, that are invisible and live in part in the mountains and in part in the deep sea, particularly in the Bermuda Triangle.

Dr. Ong's theory is different: he assumes that UFO-related phenomena are brought about by spirits or ghosts, ectoplasms or thought forms created by human beings. To corroborate his theory he refers to the research undertaken about the Poltergeist phenomenon that showed that Poltergeists are emanations from the mind of highly electric nervous adolescents. This evidence then, undertaken in the 1970's at Stanford University on the young Uri Geller, cannot seriously be contradicted.

The logical parallel that Dr. Ong forwards here is highly original, and it is highly probable. It is perhaps more probable than not that those phenomena are actually part of our own sphere while they may be part of other life dimensions or vibrational fields that we do not fully understand yet. Other evidence that corroborates Dr. Ong's view of the UFO problem is the UFO grid, an astounding phenomenon that relates to the universal *Feng Shui Dodecahedron World Grid* the existence of which Dr. Ong proves with convincing arguments and factual backup.

This universal bioenergetic world grid has its roots in ancient times and was known in Greek philosophy as the *12 Pieces of Skin* or in Russian esoteric philosophy as the *Dodecahedron Crystal*. The factual evidence produced by  the author that relates in detail to various UFO sightings and reports from reputed sources is dumbfounding and seems to prove the fact that all these phenomena feed upon earth energies or telluric energies that are emanating from underground water, *ley lines* or monuments that have perhaps been erected for this purpose by the giants or angels, such as the pyramids, Stonehenge and others. The amazing fact is namely that all these monuments can be shown to have been built exactly on the crossing of two intersecting ley lines, which are telluric energy paths in the earth's aura.

Furthermore, Dr. Ong examines the bird migration phenomenon and finds that it corroborates the evidence forwarded for the existence of the world grid – the fact is that the birds more or less follow those lines and that the energy that emanates from them serves the birds as a navigation help.

The author then refers back to Oriental sources of knowledge and examines the Middle East concept of dragon energy forces and the basics of the Chinese *Water Dragon Classic* in order to prepare the reader for a still larger perspective: the links that exist between the global knowledge of telluric lines, the Feng Shui concept of fault lines, the link between Feng Shui lines and megaliths.

Further, the author makes Feng Shui to be understood from a perspective of the Western scientist who applies his own known concepts to the ones that are not yet officially integrated into Western science. For this purpose, Dr. Ong puts forward evidence that Feng Shui actually establishes a more unified and harmonic form of perception that cares for a balance between the right and the left brain hemispheres.

At the end of his extensive study, Dr. Ong further deepens his UFO theory and asks if we are not dealing here with living beings instead of machines? He shows the links that exist between ley lines and the appearance of those spirit-UFO's which can be said to represent an extraordinary evolution of the whole of UFO research.

The study then proceeds to backup the findings with other evidence from such different sources as the *March of the Lemmings*, the 12-years *Sun Spot Cycle* and the deeper esoteric meaning of the word Feng Shui and the other Chinese words ch'i and sha.

My impression about this book is very positive. The reader of this review may have got the feeling that Dr. Ong's study is too vast and therefore difficult to understand or to read. The contrary is the case! Since years I have not got a book in my hands that I devoured with similar amazement, curiosity and inner tension. It was actually more like reading a fascinating novel than a dry scientific study. This is to say that Dr. Ong knows to write which adds to his amazing synthetic thinking capacity. The book has enriched me and given me encouragement to proceed with my own research on the still little understood functionality of the cosmic life energy. Many of the sources cited and explained by Dr. Ong actually corroborate my own findings about the cosmic energy.

There is only one negative point to mention about this book. It was obviously difficult for Dr. Ong to find a publisher in his home country Malaysia that provides him with a correct editorial work and spell-checking. The book definitely does not comply with Western or international standards of editorial work.

Evelyn Lip

The Design of Feng Shui Logos, Trademarks & Signboards

Singapore: Prentice Hall, 1995

The Design & Feng Shui of Logos, Trademarks and Signboards by Evelyn Lip is a really useful booklet about the successful design of company logos and design features used in stationary or advertisement. As the author rightly remarks in the Preface of the book, there are indeed very few books on this important subject. In practice, many company owners or even large corporations leave it completely in the hands of designers to brand their company image. There are of course good reasons for doing this, for not everyone has got aesthetic sense or the capacity to draw a logo free-hand or on the computer.

On the other hand, nobody else than yourself, founder, owner, CEO or General Manager of an organization can feel what you want to convey to the general public with the image you are going to present as a mark and symbol for what you are doing or intending to do.

Designers are now slowly catching up and some of them have implemented principles of Feng Shui in their designs. But those designers are still the exception, at least in the West. There is not only the aesthetic side in design. Feng

Shui namely teaches us that in every design or symbol there is also an energetic impact. Actually, Carl Jung has found the same to be true in his research on symbols and archetypes. Into every symbol an energy is woven which can be used either positively or negatively, depending on the intention of the one who uses it, but depending also on his or her knowledge about the power of symbols.

Symbols are often archetypal in character and thus express content that is common to all of us, through our being connected within the universal spirit network of the collective unconscious. Feng Shui expresses this in other terms but it means the same. In the Introduction, Evelyn Lip helps raising our awareness regarding the importance of impacting on our business destiny through taking appropriate action. The booklet very carefully examines and discusses successful and less successful logos and signboards after a brief introduction into the cultural impact of logo design. Among the design principles that the author outlines and marks as more or less general Feng Shui design principles, is the principle of *less is more*. In her discussion of this principle, the author discusses the *Lufthansa* logo as an example, saying that the emblem is graphically simple yet significant and is perceived as synonymous with quality, forming the basis for a sophisticated information system.

Another principle that the author qualifies as a Feng Shui principle for logo design is the expression of unity, strength and harmony. Here, as an example, the author depicts the Chase Manhattan Bank logo for further illustration.

The booklet is a must in the library of every logo design company and a useful item to study for every business owner. Usually what happens while reading the booklet - this is my own reading experience - is that from the quantity of logos depicted and discussed, a kind of vague feeling comes up for the logo that one searches for one's own business. In my case, the feeling I got was interestingly to abandon all my previous ideas of a logo that uses a picture related in some way to my business, and instead use a very simple text logo without any graphic.

A pity only that such a lucid book has got such a mediocre cover design! I have flattened the awful color setup and the *even more dreadful jpg-compression* used by amazon.com, to reproduce the cover here in simple b/w. It's beyond my grasp how an author who writes a book about design can let all of this happen! It's perhaps just another indication of many I have collected over the years of how little support and empowerment authors get from the publishing industry.

Karen Kingston

Creating Sacred Space with Feng Shui

New York: Broadway Books, 1997

Creating Sacred Space With Feng Shui is much more than a Feng Shui book. It is a whole life philosophy, and written from a strongly personal yet nonetheless verifiable perspective. This book is not a boring manual of Feng Shui. It is a most original piece of writing, unheard-of in some way.

Karen Kingston's unique talent is an inborn and absolutely stunning natural sense for the higher dimensions of existence, for all that is invisible to our physical eyes and undetectable for our five senses. The novice reader may be astonished about the authority that this text reveals and the power of the author's approach to Feng Shui that is the pragmatic and direct approach of an experienced practitioner.

There is something almost magic about this book. I have been immersed in it twice, this book being one of the very few books I have read more than once. The charm and the intuitive wisdom that the author transmits in an invisible way has kept me fascinated until the last word - and this equally during my second study.

But this book is not simply one of those poetic writings that elaborate a magic view of life. It is that also, but it is much more than that. Behind the beautiful appearance and the refined language is hidden a hard core manual that is truly scientific - in the sense of a higher and holistic form of science. Of course, the representatives of today's reductionist modern science would question most of Karen Kingston's scientific concepts. But this argument is true for almost all publications about Feng Shui.

What Karen Kingston does is exactly to go beyond the limits of a Cartesian science that is based upon wrong premises about life. To call Karen Kingston's approach to life or spirituality animistic, an argument that has been put forth even against such enlightened spirits as Johann Wolfgang von Goethe, Emanuel Swedenborg or Carl von Reichenbach would totally disregard the deep and intuitive truth that is at the basis of this holistic life philosophy.

Karen Kingston who is married with a Balinese and who lives several months every year in Bali, gives pertinent information in her book about the ways that Balinese use Feng Shui or Space Clearing. In my own experience, there are in Bali actually two levels of handling spiritual wisdom, a professional level - if I may say so - and a popular or intuitive level. The professional level is since many generations in the hands of the first caste, and especially the Pedandas (Hindu priests).

Here, we encounter a highly sophisticated and informed way of handling spiritual information that is so complex and so deep that most Westerners usually only shake their heads when they first hear about it. On the other hand there are 'the people in the street' who, in Bali, it seems, are also wiser as anywhere else in the world. For they, too, have this knowledge, only in a more intuitive and less literary form.

Having lived and worked myself in Bali for several years, I understand Karen Kingston's natural affinity with Bali and the Balinese. I could not imagine where else somebody like her could live. Following the book's advice and information about Balinese temple bells that are wonderful for clearing space with sound, I have myself acquired a Balinese temple bell, space-cleared my villa in Bali with it and can fully confirm Karen Kingston's detailed description of all the benefits that the sound of these bells has on the whole of our organism.

It all sounds like a miracle but I am convinced that it is all but magic and we will fully understand it once we know more about resonance phenomena, the complex influences that sound and vibrations have on our aura, on all our etheric bodies. The ancient musical healers knew all about it as the myth of Orpheus reveals. Compared to an approach to Feng Shui such as the one of Lillian Too, Karen Kingston seems to go much too far in her definition of Feng Shui. But it would be reductionist to trace such kind of borders.

Anyway, there is no doubt that Karen Kingston's approach to Feng Shui is one of the most original ones existing presently on earth.

Lillian Too

Feng Shui

Kuala Lumpur: Konsep Books, 1994 (Book)
Kuala Lumpur: Konsep Books, 2003 (CD)

Lillian Too is well-known in Asia. Her books are to be found in almost every major bookstore. The old science of *Feng Shui* has great appeal for the Asian public. Lillian Too is best suited to engage in this science. She used to be a very successful business woman who was, before she settled to write book after book about Feng Shui, President & CEO of a large bank in Kuala Lumpur, Malaysia. Her Chinese family tradition, too, may have contributed in her interest in this traditional Chinese science. But her success is certainly also due to the fact that business people in Asia have a high regard for Feng Shui, which is for the most part ignored by Western business people, even those living and working in Asia.

In fact, it has taken generations until the West awoke from its materialistic trance and inquired if there was not something more subtle in life than what the five senses can grasp. And now, paradoxically, it is for the most part the Oriental approach to geomancy, Feng Shui, that is becoming popular, and not our own Western geomancy tradition which is as erudite as its Oriental counterpart.

Dr. Ong, in the book that I discussed earlier in this book review, reminds us that the Druid sages were reported to be able to ride on the subtle energies so that they could fly in the air without any device other than the forces of the sun and the moon that they knew to activate for their purposes. Of course, the modern reader is less interested in these stories than in receiving practical and down-to-earth advice how he can improve luck, health, happiness and wealth in their lives. And this pragmatic kind of approach is exactly the tenor of most Feng Shui books that are published in Asia. Lillian Too gives such advice and she does it in a straightforward manner that is exemplary. Her working together with renowned Feng Shui master Yap Cheng Hai led to a fruitful collaboration that the books brilliantly testify.

Lillian Too defines Feng Shui as *The Art of Living in Harmony with the Land*, the deeper wisdom that teaches us the power of being in harmony with all our surroundings. A wisdom, to repeat it, that was taught equally in the West, all through our history, that however was disregarded by the power institutions like the Church and later the nation states that were not interested in the subtle truth of life.

The book proceeds teaching Feng Shui in twenty-three chapters. The style of the book is very methodic and it presents knowledge in the traditional, deductive way. Whereas now in the West the trend goes more in the direction of presenting new and unusual knowledge inductively or empirically. This rather traditional and academic way of presenting content may at times contribute to a somewhat boring reading experience. However, the author knows to act counter to this danger by giving many examples of her practice, telling anecdotes or referring to the writings of the old masters that she appears to have studied extensively.

The book is certainly a very good starting point for more in-depth Feng Shui studies and practice, and so much the more for readers who are trained as scientists or in any other traditional, methodic way.

Man-Ho Kwok

The Feng Shui Kit

London: Piatkus, 1995

The Feng Shui Kit was the very first book I found on the subject of Feng Shui, ten years ago. And it is more than a book! On the back of the cover I had written *Bookstore Hyatt Regency*, Surabaya, Indonesia, April 3, 1996.

This Kit had a very strong impact on me, not only because it had been the first publication that got in my hands about the fascinating perennial science of Feng Shui, but because it was an item to play with, something I could become active with right away. For there is a *Lo Pan* delivered in the kit, which is a Feng Shui Compass, though from plastic. But it contains the basic signs and directions, and in the book you can look up the explanation.

To give a practical example. The first thing to do is to adjust a wheel on the compass the way that the red-white compass needle is exactly in the middle of two little red points. The direction here indicated is South. The compass actually possesses two wheels that you can turn. The greater wheel that is behind contains your personal animals in Chinese astrology. Mine being *Ram*, I turn this black wheel now while holding the red one, to which a compass is fixed, so that Ram points to the direction, place or location I want to assess the Feng Shui of.

Before I demonstrate how the kit works, you need to know which elements are contained in every single reading. You have to follow through since if you leave out one, you will get a distorted answer. These elements are represented by *five rings* on the Lo Pan:

– 1st Ring: I Ching Hexagram
– 2nd Ring: Direction
– 3rd Ring: Element
– 4th Ring: Yin/Yang
– 5th Ring: Animal

So let's do a reading right here. I am sitting now here at my desk, in a newly furnished room that since I have entered this rented house, I have not used yet. My face points to the entry door and it is this direction that I would like to assess. So I adjust my compass so that the red needle is between the two red dots and thus points South. I look up the direction I want to assess and get the following information:

First ring (I Ching sign): Chen

Now the book tells me that I need to know the other trigram so that I can look up the complete hexagram. In order to calculate the other trigram, the Kit explains, I need to make the following calculation (for males) which is called Your personal Pa Tzu Compass:

> **Man-Ho Kwok**
> Subtract the last two numbers of my year of birth from 100 and divide by 9.

My year of birth being 1955, I arrive at the number 5. Now I search for my personal compass and do not find it. I look further and see that people with compass number 5 should use compass number 2 if they are males. The trigram associated with compass number 2 being K'un I look up the combination of Chen (thunder) and K'un (earth) and arrive at the I Ching hexagram *24. Return*. The explanation reads at follows:

> **Man-Ho Kwok**
> This hexagram indicates a period of growth after a period of disorder and disintegration. It is a positive hexagram and is full of possibilities. It is time for a fresh start so enlist the help of others, but do not try to rush change as it will take a natural course.

Second Ring (Direction): East

Very unfavorable for compass No. 2. Since the entry of my house points to that direction, this could signify that I should not go out much, which actually I don't do. Culturally and socially, there is as good as nothing attractive for me here on *Lombok* island, except some nice beaches that, however, I do not frequent often since I am working very intensely on around forty publishing projects. On the other hand, the house I got here, a huge villa owned by an Arab merchant, is unique and represents in many ways what I have always dreamt of. So I really enjoy to be here and to have such a wonderful place for my work, too.

Third Ring (Element): Earth

The book's explanation suggests to introduce plants or wooden products in that area. When I watch out of my entry door that my desk faces, I see the plants, bushes and trees of my front yard. So this is well adjusted. However, my personal element according to Compass No. 2 being Wood and according to the destructive cycle Wood destroys Earth, this reading is also partly negative.

Fourth Ring (Yin/Yang): Yang

The fourth ring indicating Yang, I have to look up the seventh ring, too, which is related to my personal animal (Ram) and that indicates Yin. Yin and Yang are mutually complementary and balance each other out. So this reading is positive.

Fifth Ring (Animal): Dragon

This ring has to bee seen in relation to the last ring which indicates the personal animal. As to the combination Dragon-Ram, the book explains:

Man-Ho Kwok
> In this combination there is a temptation to let your imagination run free with design and color, but it may not always please others. You should try to be more open to advise.

This is true! I have behind me a phase of designing web pages and I wanted to impress my visitors with strong and bold colors. There was much red and orange. Eventually, after going on months and months that way, I got feedback from a good and close friend who told me he felt irritated and confused when looking at my pages. There was too much and too many colors so that at the end, he confessed, he turned away since he did not catch the message. What a change to have got his advise!

When I look back now at my former period, I cannot understand myself anymore. How could I have liked that? But that's the way we change, *when* we change. That I should be more open for advise is equally true. Had I been more open to follow the advise of experts I would have avoided to add many unnecessary features in my web design. It would have saved me time and money to have followed this advise.

I admit that the readings you get using this Feng Shui Kit are somewhat superficial; but, of course, there are more erudite publications. Only, you have to have the time to really digest them and apply them in your life. To just fill your bookshelf with would be another waste of resources. So I take the essence out of the Kit's reading and complete it using my own intuition. That's a good combination, anyway!

Nancilee Wydra

The Feng: The Book of Cures

Lincolnwood: Contemporary Books, 1996

The Book of Cures is one of the practical Feng Shui books. The book, while it is compact in size, is more like a handbook or manual. And as it is often the case with such kind of books where the overwhelming part consists of looking up things (instead of reading the whole), the author is conscious about the fact that the reader needs some introduction in the general principles underlying the advice presented in the manual, before they can successfully apply that advice. And this introduction is excellent! It is among the best for those who desire a very short, very compact and very easy-to-read introduction into the highly complex science of Feng Shui.

I would go as far as saying that already the introduction (Part I) is worth buying this book. I myself do not need the practical advise or cures described in Part II, since I practice intuitive Feng Shui since many years, and even before I knew about this science.

Some glimpses, however, that I made into the Cures part of the book showed me that the advise is appropriate and written in a comprehensive manner that is easy to understand. I therefore concentrate upon this little jewel that the author may herself not consider as the major contribution this booklet makes: the *introduction*. It is divided into six chapters:

- Power of Place
- What is Feng Shui?
- Schools of Feng Shui
- The Five Elements
- The Pyramid School's Ba-Gua
- The Senses

This introductory part of the book is the most necessary to read for any uninitiated Western reader. Let me cite the way the author opens her book:

> **Nancilee Wydra**
> I can recall how places felt to me as a child. On my way to school each morning my girlfriends and I walked past our neighborhood's haunted house. The early morning sun peeping up from behind this home's conical roof created an ominous shadow on the cracked cement sidewalk.

Almost everybody can report similar feelings and observations from our childhood. But how many of us have noted them or inquire into their validity once being grown up? I have myself gone all the way from a materialistic worldview which was shaped during all those years of school terror until my today's open, holistic, dynamic and complexity-affirming worldview. It seems that only poets and very strong individuals are able to have their childhood intuition prevail over that cruel brainwashing school has done to most of us. I have fortunately gained it back by recovering and healing my inner child during a two-years hypnotherapy some years ago. Today I cannot understand how one can live with the dangerously reductionist worldview that is still the reigning consciousness paradigm of the majority of people in our times.

As Nancilee Wydra suggests, we can in fact dive into our childhood memories again and see what our knowledge was regarding power and places or energies that radiate from places or that is kept attached to objects. Her approach is interesting also from another point of view. She questioned in her book the immediate applicability of ancient concepts or concepts pertaining to certain cultures, in other cultures. For example, she cites ways of ill-interpreting Feng Shui rules because those rules fit a certain culture (the Chinese culture) with a certain set of behavior. She therefore suggests to abstract or extrapolate the underlying principle of every rule and then apply the extracted principle to the problem, and not the literal rule. This seems to be a wise approach, indeed!

I know from experience that it is exactly the lack of this capacity to abstract the general principle from ancient rules that make people today turn away from them. Dr. Joseph Murphy found the same to be true regarding the Bible truths. Applied literally, many of those truths would be absurd in our times or they would hurt people more than healing them. It is only through interpreting the ancient texts in the light of our today's psychological understanding that we can see the true and deep meaning of what is written. To develop this understanding means to work with our intuitive and synthetic mind, using more of our right brain hemisphere's associative thinking capacity, than applying the strict logic ability of the left brain. This book is a step in the right direction also in this sense. After all, I can recommend this book especially to those who are new to the subject of Feng Shui, who are curious to learn the basics in a comprehensive way that takes a minimum of time.

Richard Craze

The Feng: Feng Shui Book and Card Pack

London: Thorsons, 1997

Feng Shui Cards are a beautiful item that convinces through *perfect harmony* between content and design. I have seldom seen a book published with such a wonderful interplay of colors and design elements. The idea is fantastic, too. You just go in a corner of your house or room, mix the cards in your hand, meditate a moment, pull out a card blindly and read it. It is that way, entirely by intuition, that the Feng Shui is assessed.

This intuitive approach to Feng Shui is so much the more satisfying for the beginner who has little knowledge yet of the intricacies and complexities of Compass Feng Shui.

On the hand, of course, one must have a feeling for that magic reality – and some people are just not to turn on that way. So this booklet is probably not a mass selling item. So much the more I compliment the courage and aesthetic sense of authors and publisher!

After a short introduction into the principles of Feng Shui, the booklet explains the approach to Feng Shui that is the most practiced now in the Western world: *Pah Kwa Feng*

Shui, which uses a fixed arrangement of trigrams, derived from the I Ching, that are positioned in form of an octahedron and that overlays the map of a room or house on it. Logically, eight different sections will show up which are called *The Eight Enrichments*. Accordingly, for each of these eight main sections, the booklet shows Eight Remedies in case the card indicates some form of negative energy (the ch'i may for example be unpredictable or overpowering in one section).

Finally, an assessment of one's whole property according to the *Five Elements* is indicated before the booklet explains every card in detail.

The 32 cards themselves are divided in four different suites, according to the Four Animals that traditional Feng Shui teaching situates around every house or property: the Red Phoenix (South), the White Tiger (West), the Black Tortoise (North) and the Green Dragon (East).

To end this review of a short but wonderful booklet, let me cite the first sentence of the Introduction because this single sentence reveals the unique and powerfully holistic focus of the booklet:

> **Richard Craze**
> Taoism, which is the ancient religion of China, holds that what is, is. Unlike Christians, who believe in the heavenly paradise of an afterlife, or Hindus with their vast array of gods, or even the Buddhists' belief that all life is suffering and the only reward is in Nirvana, Taoists regard now as important and believe that there are no gods, heavens, or future paradises. Their heaven is order, harmony, balance and jen - love of life.

Apart from what this set represents in knowledge and practical value, it is also a wonderful gift item!

BIBLIOGRAPHY

Contextual Bibliography

Arntz, William & Chasse, Betsy
What the Bleep Do We Know
20th Century Fox, 2005 (DVD)

Down The Rabbit Hole Quantum Edition
20th Century Fox, 2006 (3 DVD Set)

Bleep
An der Schnittstelle von Spiritualität und Wissenschaft
Verblüffende Erkenntnisse und Anstösse zum Weiterdenken
Berlin: Vak Verlag, 2007

Arroyo, Stephen
Astrology, Karma & Transformation
The Inner Dimensions of the Birth Chart
Sebastopol, CA: CRSC Publications, 1978

Astrologie, Karma und Transformation
Die Chancen schwieriger Aspekte
Frankfurt/M: Heyne Verlag, 1998

Relationships and Life Cycles
Astrological Patterns of Personal Experience
Sebastopol, CA: CRCS Publications, 1993

Handbuch der Horoskop-Deutung
Berlin: Rowohlt, 1999

Bachelard, Gaston
The Poetics of Reverie
Translated by Daniel Russell
Boston: Beacon Press, 1971

Poetik des Raumes
Frankfurt/M: Fischer Verlag, 2001

Balter, Michael
The Goddess and the Bull
Catalhoyuk, An Archaeological Journey
to the Dawn of Civilization
New York: Free Press, 2006

Bandler, Richard
Get the Life You Want
The Secrets to Quick and Lasting Life Change
With Neuro-Linguistic Programming
Deerfield Beach, Fl: HCI, 2008

Blofeld, J.
The Book of Changes
A New Translation of the Ancient Chinese I Ching
New York: E.P. Dutton, 1965

Blum, Ralph H. & Laughan, Susan
The Healing Runes
Tools for the Recovery of Body, Mind, Heart & Soul
New York: St. Martin's Press, 1995

Bohm, David
Wholeness and the Implicate Order
London: Routledge, 2002

Die implizite Ordnung
Grundlagen eines dynamischen Holismus
München: Goldmann Wilhelm, 1989

Thought as a System
London: Routledge, 1994

Quantum Theory
London: Dover Publications, 1989

La plénitude de l'univers
Paris: Rocher, 1992

Branden, Nathaniel

How to Raise Your Self-Esteem
New York: Bantam, 1987

Die 6 Säulen des Selbstwertgefühls
Erfolgreich und zufrieden durch ein starkes Selbst
München: Piper Verlag, 2009

Butler-Bowden, Tom

50 Success Classics
Winning Wisdom for Work & Life From 50 Landmark Books
London: Nicholas Brealey Publishing, 2004

Boldt, Laurence G.

Zen and the Art of Making a Living
A Practical Guide to Creative Career Design
New York: Penguin Arkana, 1993

How to Find the Work You Love
New York: Penguin Arkana, 1996

Zen Soup
Tasty Morsels of Zen Wisdom From Great Minds East & West
New York: Penguin Arkana, 1997

The Tao of Abundance
Eight Ancient Principles For Abundant Living
New York: Penguin Arkana, 1999

Das Tao der Fülle
Vom Reichtum, der uns glücklich macht
Mittelberg: Joy Verlag, 2001

Campbell, Joseph

The Hero With A Thousand Faces
Princeton: Princeton University Press, 1973
(Bollingen Series XVII)
London: Orion Books, 1999

Der Heros in Tausend Gestalten
München: Insel Verlag, 2009

Occidental Mythology
Princeton: Princeton University Press, 1973
(Bollingen Series XVII)
New York: Penguin Arkana, 1991

The Masks of God
Oriental Mythology
New York: Penguin Arkana, 1992
Originally published 1962

Mythologie des Ostens
Die Masken Gottes Bd. 2
Basel: Sphinx Verlag, 1996

The Power of Myth
With Bill Moyers
ed. by Sue Flowers
New York: Anchor Books, 1988

Die Kraft der Mythen
Düsseldorf: Patmos Verlag, 2007

Capacchione, Lucia

The Power of Your Other Hand
North Hollywood, CA: Newcastle Publishing, 1988

Capra, Bernt Amadeus

Mindwalk
A Film for Passionate Thinkers
Based Upon Fritjof Capra's *The Turning Point*
New York: Triton Pictures, 1990

Capra, Fritjof

The Turning Point
Science, Society And The Rising Culture
New York: Simon & Schuster, 1987
Original Author Copyright, 1982

Wendezeit
Bausteine für ein neues Weltbild
München: Droemer Knaur, 2004

Le temps du changement
Science, société et nouvelle culture
Paris: Rocher, 1994

The Tao of Physics
An Exploration of the Parallels Between Modern
Physics and Eastern Mysticism
New York: Shambhala Publications, 2000
(New Edition) Originally published in 1975

Das Tao der Physik
Die Konvergenz von westlicher Wissenschaft und östlicher Philosophie
Neue und erweiterte Auflage
München: O.W. Barth bei Scherz, 2000
Ursprünglich erschienen 1975 bei Droemersche Verlagsanstalt
in Hamburg

Le tao de la physique
Paris: Sand & Tchou, 1994

The Web of Life
A New Scientific Understanding of Living Systems
New York: Doubleday, 1997

Lebensnetz
Ein neues Verständnis der lebendigen Welt
München: Scherz Verlag, 1999

The Hidden Connections
Integrating The Biological, Cognitive And Social Dimensions Of Life Into A Science Of Sustainability
New York: Doubleday, 2002

Verborgene Zusammenhänge
München: Scherz, 2002

Steering Business Toward Sustainability
New York: United Nations University Press, 1995

Uncommon Wisdom
Conversations with Remarkable People
New York: Bantam, 1989

The Science of Leonardo
Inside the Mind of the Great Genius of the Renaissance
New York: Anchor Books, 2008
New York: Bantam Doubleday, 2007 (First Publishing)

Chopra, Deepak

Creating Affluence
The A-to-Z Steps to a Richer Life
New York: Amber-Allen Publishing (2003)

Synchrodestiny
Discover the Power of Meaningful Coincidence to Manifest Abundance
Audio Book / CD
Niles, IL: Nightingale-Conant, 2006

The Seven Spiritual Laws of Success
A Practical Guide to the Fulfillment of Your Dreams
Audio Book / CD
New York: Amber-Allen Publishing (2002)

Die Sieben Geistigen Gesetze des Erfolgs
Berlin: Ullstein Verlag, 2004

The Spontaneous Fulfillment of Desire
Harnessing the Infinite Power of Coincidence
New York: Random House Audio, 2003

Covey, Stephen R.

The 7 Habits of Highly Effective People
Powerful Lessons in Personal Change
New York: Free Press, 2004
15th Anniversary Edition
First Published in 1989

Die 7 Wege zur Effektivität
Prinzipien für persönlichen und beruflichen Erfolg
Offenbach: Gabal Verlag, 2009

The 8th Habit
From Effectiveness to Greatness
London: Simon & Schuster, 2004

Der 8. Weg
Von der Effektivität zur wahren Grösse
6. Auflage
Offenbach: Gabal Verlag, 2006

Craze, Richard

Feng Shui
Feng Shui Book & Card Pack
London: Thorsons, 1997

De Bono, Edward

The Use of Lateral Thinking
New York: Penguin, 1967

The Mechanism of Mind
New York: Penguin, 1969

Sur/Petition
London: HarperCollins, 1993

Tactics
London: HarperCollins, 1993
First published in 1985

Taktiken und Strategien erfolgreicher Menschen
Frankfurt/M: MVG Verlag, 1995

Serious Creativity
Using the Power of Lateral Thinking to Create New Ideas
London: HarperCollins, 1996

Dürckheim, Karlfried Graf

Hara: The Vital Center of Man
Rochester: Inner Traditions, 2004

Hara
Die Erdmitte des Menschen
Neuausgabe
München: O.W. Barth bei Scherz, 2005

Zen and Us
New York: Penguin Arkana 1991

The Call for the Master
New York: Penguin Books, 1993

Absolute Living
The Otherworldly in the World and the Path to Maturity
New York: Penguin Arkana, 1992

The Way of Transformation
Daily Life as a Spiritual Exercise
London: Allen & Unwin, 1988

Der Alltag als Übung
Vom Weg der Verwandlung
Bern: Huber, 2008

The Japanese Cult of Tranquility
London: Rider, 1960

Kultur der Stille
Frankfurt/M: Weltz Verlag, 1997

Emoto, Masaru
The Hidden Messages in Water
New York: Atria Books, 2004

Die Botschaft des Wassers
Burgrain: Koha Verlag, 2008

The Secret Life of Water
New York: Atria Books, 2005

Die Heilkraft des Wassers
Burgrain: Koha Verlag, 2004

Goleman, Daniel
Emotional Intelligence
New York, Bantam Books, 1995

EQ. Emotionale Intelligenz
München: DTV Verlag, 1997

Goswami, Amit
The Self-Aware Universe
How Consciousness Creates the Material World
New York: Tarcher/Putnam, 1995

Das Bewusste Universum
Wie Bewusstsein die materielle Welt erschafft
Stuttgart: Lüchow Verlag, 2007

Greene, Liz
Astrology of Fate
York Beach, ME: Red Wheel/Weiser, 1986

Saturn
A New Look at an Old Devil
York Beach, ME: Red Wheel/Weiser, 1976

The Astrological Neptune and the Quest for Redemption
Boston: Red Wheel Weiser, 1996

The Mythic Journey
With Juliet Sharman-Burke
The Meaning of Myth as a Guide for Life
New York: Simon & Schuster (Fireside), 2000

Die Mythische Reise
Die Bedeutung der Mythen als ein Führer durch das Leben
München: Atmosphären Verlag, 2004

The Mythic Tarot
With Juliet Sharman-Burke
New York: Simon & Schuster (Fireside), 2001
Originally published in 1986

Le Tarot Mythique
Une nouvelle approche du Tarot
Paris: Solar, 1988

The Luminaries
The Psychology of the Sun and Moon in the Horoscope
With Howard Sasportas
York Beach, ME: Red Wheel/Weiser, 1992

Sonne und Mond
Die Bedeutung der grossen Lichter in der Mythologie und im Horoskop
Saarbrücken: Neue Erde/Lentz, 2000

Greer, John Michael

Earth Divination, Earth Magic
A Practical Guide to Geomancy
New York: Llewellyn Publications, 1999

Hicks, Esther and Jerry
The Amazing Power of Deliberate Intent
Living the Art of Allowing
Carlsbad, CA: Hay House, 2006

Holmes, Ernst
The Science of Mind
A Philosophy, A Faith, A Way of Life
New York: Jeremy P. Tarcher/Putnam, 1998
First Published in 1938

Houston, Jean
The Possible Human
A Course in Enhancing Your Physical, Mental, and Creative Abilities
New York: Jeremy P. Tarcher/Putnam, 1982

Huang, Alfred
The Complete I Ching
The Definite Translation from Taoist Master Alfred Huang
Rochester, NY: Inner Traditions, 1998

Hunt, Valerie
Infinite Mind
Science of the Human Vibrations of Consciousness
Malibu, CA: Malibu Publishing, 2000

Huxley, Aldous
The Doors of Perception and Heaven and Hell
London: HarperCollins (Flamingo), 1994
(originally published in 1954)

The Perennial Philosophy
San Francisco: Harper & Row, 1970

Jackson, Nigel
The Rune Mysteries
With Silver RavenWolf
St. Paul, Minn.: Llewellyn Publications, 2000

Jung, Carl Gustav
Archetypen
München: DTV Verlag, 2001

Archetypes of the Collective Unconscious
in: The Basic Writings of C.G. Jung
New York: The Modern Library, 1959, 358-407

Collected Works
New York, 1959

Dialectique du moi et de l'inconscient
Paris, Gallimard, 1991

On the Nature of the Psyche
in: The Basic Writings of C.G. Jung
New York: The Modern Library, 1959, 47-133

Psychological Types
Collected Writings, Vol. 6
Princeton: Princeton University Press, 1971

Psychologie und Religion
München: DTV Verlag, 2001

Psychology and Religion
in: The Basic Writings of C.G. Jung
New York: The Modern Library, 1959, 582-655

Religious and Psychological Problems of Alchemy
in: The Basic Writings of C.G. Jung
New York: The Modern Library, 1959, 537-581

Symbol und Libido
Freiburg: Walter Verlag, 1987

Synchronizität, Akausalität und Okkultismus
Frankfurt/M: DTV, 2001

The Basic Writings of C.G. Jung
New York: The Modern Library, 1959

The Development of Personality
Collected Writings, Vol. 17
Princeton: Princeton University Press, 1954

The Meaning and Significance of Dreams
Boston: Sigo Press, 1991

The Myth of the Divine Child
in: Essays on A Science of Mythology
Princeton, N.J.: Princeton University Press Bollingen
Series XXII, 1969. (With Karl Kerenyi)

Traum und Traumdeutung
München: DTV Verlag, 2001

Two Essays on Analytical Psychology
Collected Writings, Vol. 7
Princeton: Princeton University Press, 1972
First published by Routledge & Kegan Paul, Ltd., 1953

Zur Psychologie westlicher und östlicher Religion
Fünfte Auflage
Olten: Walter Verlag, 1988

Karagulla, Shafica

The Chakras
Correlations between Medical Science and Clairvoyant Observation
With Dora van Gelder Kunz
Wheaton: Quest Books, 1989

Die Chakras und die feinstofflichen Körper des Menschen
Mit Dora van Gelder-Kunz
Grafing: Aquamarin Verlag, 1994

Kiang, Kok Kok

The I Ching
An Illustrated Guide to the Chinese Art of Divination
Singapore: Asiapac, 1993

Kingston, Karen

Creating Sacred Space With Feng Shui
New York: Broadway Books, 1997

Krishnamurti, J.

Freedom From The Known
San Francisco: Harper & Row, 1969

The First and Last Freedom
San Francisco: Harper & Row, 1975

Education and the Significance of Life
London: Victor Gollancz, 1978

Commentaries on Living
First Series
London: Victor Gollancz, 1985
Commentaries on Living
Second Series
London: Victor Gollancz, 1986

Krishnamurti's Journal
London: Victor Gollancz, 1987

Krishnamurti's Notebook
London: Victor Gollancz, 1986

Beyond Violence
London: Victor Gollancz, 1985

Beginnings of Learning
New York: Penguin, 1986

The Penguin Krishnamurti Reader
New York: Penguin, 1987

On God
San Francisco: Harper & Row, 1992

On Fear
San Francisco: Harper & Row, 1995

The Essential Krishnamurti
San Francisco: Harper & Row, 1996

The Ending of Time
With Dr. David Bohm
San Francisco: Harper & Row, 1985

Kwok, Man-Ho

The Feng Shui Kit
London: Piatkus, 1995

Lakhovsky, Georges

La Science et le Bonheur
Longévité et Immortalité par les Vibrations
Paris: Gauthier–Villars, 1930

Le Secret de la Vie
Paris: Gauthier–Villars, 1929

Secret of Life
New York: Kessinger Publishing, 2003

L'étiologie du Cancer
Paris: Gauthier–Villars, 1929

L'Universion
Paris: Gauthier–Villars, 1927

Leadbeater, Charles Webster

Astral Plane
Its Scenery, Inhabitants and Phenomena
Kessinger Publishing Reprint Edition, 1997

Dreams
What they Are and How they are Caused
London: Theosophical Publishing Society, 1903
Kessinger Publishing Reprint Edition, 1998

The Inner Life
Chicago: The Rajput Press, 1911
Kessinger Publishing

Leonard, George, Murphy, Michael

The Live We Are Given
A Long Term Program for Realizing the
Potential of Body, Mind, Heart and Soul
New York: Jeremy P. Tarcher/Putnam, 1984

Liedloff, Jean

Continuum Concept
In Search of Happiness Lost
New York: Perseus Books, 1986
First published in 1977

Auf der Suche nach dem verlorenen Glück
Gegen die Zerstörung der Glücksfähigkeit in der frühen Kindheit
München: C.H. Beck Verlag, 2006

Lip, Evelyn

The Design & Feng Shui of Logos, Trademarks and Signboards
Singapore: Prentice Hall, 1995

Long, Max *Freedom*

The Secret Science at Work
The Huna Method as a Way of Life

Marina del Rey: De Vorss Publications, 1995
Originally published in 1953

Geheimes Wissen hinter Wundern
Die Entdeckung der HUNA-Lehre
Darmstadt: Schirner Verlag, 2006

Growing Into Light
A Personal Guide to Practicing the Huna Method,
Marina del Rey: De Vorss Publications, 1955

Lowen, Alexander
Angst vor dem Leben
Über den Ursprung seelischen Leides und den Weg
zu einem reicheren Dasein
München: Goldmann Wilhelm, 1989

Bioenergetics
New York: Coward, McGoegham 1975

Bioenergetik
Therapie der Seele durch Arbeit mit dem Körper
Berlin: Rowohlt, 2008

Depression and the Body
The Biological Basis of Faith and Reality
New York: Penguin, 1992

Fear of Life
New York: Bioenergetic Press, 2003

Honoring the Body
The Autobiography of Alexander Lowen
New York: Bioenergetic Press, 2004

Joy
The Surrender to the Body and to Life
New York: Penguin, 1995

Love and Orgasm
New York: Macmillan, 1965

Love, Sex and Your Heart
New York: Bioenergetics Press, 2004

Narcissism: Denial of the True Self
New York: Macmillan, Collier Books, 1983

Narzissmus
Die Verleugnung des wahren Selbst
München: Goldmann Wilhelm, 1992

Pleasure: A Creative Approach to Life
New York: Bioenergetics Press, 2004
First published in 1970

The Language of the Body
Physical Dynamics of Character Structure
New York: Bioenergetics Press, 2006

Maharshi, Ramana

The Collected Works of Ramana Maharshi
New York: Sri Ramanasramam, 2002

The Essential Teachings of Ramana Maharshi
A Visual Journey
New York: Inner Directions Publishing, 2002
by Matthew Greenblad

Sei was du bist!
München: O.W. Barth, 2001

Nan Yar? Wer bin ich?
München: Kamphausen, 2002

Malinowski, Bronislaw

Crime und Custom in Savage Society
London: Kegan, 1926

Sex and Repression in Savage Society
London: Kegan, 1927

The Sexual Life of Savages in North West Melanesia
New York: Halycon House, 1929

Das Geschlechtsleben der Wilden in Nordwest-Melanesien
Liebe, Ehe und Familienleben bei den Eingeborenen der
Trobriand Inseln, Britisch-Neuguinea
Eschborn: Klotz Verlag, 2005

McKenna, Terence
The Archaic Revival
San Francisco: Harper & Row, 1992

Food of The Gods
A Radical History of Plants, Drugs and Human Evolution
London: Rider, 1992

Die Speisen der Götter
Berlin: Synergia/Syntropia, 1996

The Invisible Landscape
Mind Hallucinogens and the I Ching
New York: HarperCollins, 1993
(With Dennis McKenna)

True Hallucinations
Being the Account of the Author's Extraordinary
Adventures in the Devil's Paradise
New York: Fine Communications, 1998

McNiff, Shaun
Art as Medicine
Boston: Shambhala, 1992

Art as Therapy
Creating a Therapy of the Imagination
Boston/London: Shambhala, 1992

Trust the Process
An Artist's Guide to Letting Go
New York: Shambhala Publications, 1998

Miller, Mary & Taube, Karl

An Illustrated Dictionary of the Gods and Symbols of Ancient Mexico and the Maya
London: Thames & Hudson, 1993

Moore, Thomas

Care of the Soul
A Guide for Cultivating Depth and Sacredness in Everyday Life
New York: Harper & Collins, 1994

Die Seele Lieben
Tiefe und Spiritualität im täglichen Leben
München: Droemer Knaur, 1995

Murphy, Joseph

The Power of Your Subconscious Mind
West Nyack, N.Y.: Parker, 1981, N.Y.: Bantam, 1982
Originally published in 1962

Die Macht Ihres Unterbewusstseins
München: Hugendubel, 2000

La puissance de votre subconscient
Genève: Ramón Keller, 1967

The Miracle of Mind Dynamics
New York: Prentice Hall, 1964

Miracle Power for Infinite Riches
West Nyack, N.Y.: Parker, 1972

The Amazing Laws of Cosmic Mind Power
West Nyack, N.Y.: Parker, 1973

Secrets of the I Ching
West Nyack, N.Y.: Parker, 1970

Think Yourself Rich
Use the Power of Your Subconscious Mind to Find True Wealth
Revised by Ian D. McMahan, Ph.D.
Paramus, NJ: Reward Books, 2001

Das Erfolgsbuch
Wie sie alles im Leben erreichen können
Hamburg: Heyne Verlag, 2002

Wahrheiten die ihr Leben verändern
Dr. Joseph Murphys Vermächtnis
München: Hugendubel, 1996

Murphy, Michael

The Future of the Body
Explorations into the Further Evolution of Human Nature
New York: Jeremy P. Tarcher/Putnam, 1992

Der Quanten-Mensch
München: Ludwig Verlag, 1996

Myers, Tony Pearce

The Soul of Creativity
Insights into the Creative Process
Novato, CA: New World Library, 1999

Narby, Jeremy

The Cosmic Serpent
DNA and the Origins of Knowledge
New York: J. P. Tarcher, 1999

Die Kosmische Schlange
Auf den Pfaden der Schamanen zu den Ursprüngen modernen Wissens
Stuttgart: Klett-Cotta, 2007

Nau, Erika

Self-Awareness Through Huna
Virginia Beach: Donning, 1981

Selbstbewusst durch Huna
Die magische Weisheit Hawaiis
2. Auflage
Basel: Sphinx Verlag, 1989

Ni, Hua-Ching

I Ching
The Book of Changes and the Unchanging Truth
2nd edition
Santa Barbara: Seven Star Communications, 1999

Esoteric Tao The Ching
The Shrine of the Eternal Breath of Tao
Santa Monica: College of Tao and Traditional
Chinese Healing, 1992

The Complete Works of Lao Tzu
Tao The Ching & Hua Hu Ching
Translation and Elucidation by Hua-Ching Ni
Santa Monica: Seven Star Communications, 1995

Nichols, Sallie

Jung and Tarot: An Archetypal Journey
New York: Red Wheel/Weiser, 1986

Die Psychologie des Tarot
Interlaken: Ansata Verlag, 1996

Ong, Hean-Tatt
Amazing Scientific Basis of Feng-Shui
Kuala Lumpur: Eastern Dragon Press, 1997

Ostrander, Sheila & Schroeder, Lynn
Superlearning 2000
New York: Delacorte Press, 1994

Superlearning
Die revolutionäre Lernmethode
München: Scherz Verlag, 1979

Supermemory
New York: Carroll & Graf, 1991

SuperMemory
Der Weg zum optimalen Gedächtnis
München: Goldmann, 1996

Radin, Dean
The Conscious Universe
The Scientific Truth of Psychic Phenomena
San Francisco: Harper & Row, 1997

Entangled Minds
Extrasensory Experiences in a Quantum Reality
New York: Paraview Pocket Books, 2006

Reich, Wilhelm
*A Review of the Theories, dating from The 17th Century,
on the Origin of Organic Life, by Arthur Hahn, Literature Assistant at the
Institut für Sexualökonomische Lebensforschung, Biologisches
Laboratorium, Oslo, 1938*
©1979 by Mary Boyd Higgins as Director of the Wilhelm Reich Infant
Trust (XEROX Copy from the Wilhelm Reich Museum)

CORE (Cosmic Orgone Engineering)
Part I, Space Ships, DOR and DROUGHT
©1984, Orgone Institute Press

Der Einbruch der sexuellen Zwangsmoral
Frankfurt/M: Fischer, 1981

Die Entdeckung des Orgons II
Der Krebs
Frankfurt/M: Fischer, 1981
Köln: Kiepenheuer & Witsch, 1984

Die Funktion des Orgasmus
Sexualökonomische Grundprobleme der biologischen Energie
Köln: Kiepenheuer & Witsch, 1987

Die Massenpsychologie des Faschismus
Frankfurt/M: Fischer, 1974

Die sexuelle Revolution
Frankfurt/M: Fischer, 1966

Early Writings 1
New York: Farrar, Straus & Giroux, 1975

Ether, God & Devil & Cosmic Superimposition
New York: Farrar, Straus & Giroux, 1972
Originally published in 1949

Frühe Schriften 1
Aus den Jahren 1920–1925
Frankfurt/M: Fischer, 1983

Frühe Schriften 2
Genitalität in der Theorie und Therapie der Neurose
Frankfurt/M: Fischer, 1985

Genitality in the Theory and Therapy of Neurosis
©1980 by Mary Boyd Higgins as Director
of the Wilhelm Reich Infant Trust

Leidenschaften der Jugend
Köln: Kiepenheuer & Witsch, 1984

L'irruption de la morale sexuelle
Paris: Payot, 1972

Menschen im Staat
Frankfurt/M: Nexus, 1982

People in Trouble
©1974 by Mary Boyd Higgins as Director
of the Wilhelm Reich Infant Trust

Record of a Friendship
The Correspondence of Wilhelm Reich and A. S. Neill
New York, Farrar, Straus & Giroux, 1981

Selected Writings
An Introduction to Orgonomy
New York: Farrar, Straus & Giroux, 1973

The Bioelectrical Investigation of Sexuality and Anxiety
New York: Farrar, Straus & Giroux, 1983
Originally published in 1935

The Bion Experiments
reprinted in *Selected Writings*
New York: Farrar, Straus & Giroux, 1973

The Cancer Biopathy (The Orgone, Vol. 2)
New York: Farrar, Straus & Giroux, 1973

The Function of the Orgasm (The Orgone, Vol. 1)
Orgone Institute Press, New York, 1942

The Invasion of Compulsory Sex Morality
New York: Farrar, Straus & Giroux, 1971
Originally published in 1932

The Leukemia Problem: Approach
©1951, Orgone Institute Press
Copyright Renewed 1979
XEROX Copy from the Wilhelm Reich Museum

The Mass Psychology of Fascism
New York: Farrar, Straus & Giroux, 1970
Originally published in 1933

The Orgone Energy Accumulator
Its Scientific and Medical Use
©1951, 1979, Orgone Institute Press
XEROX Copy from the Wilhelm Reich Museum

The Schizophrenic Split
©1945, 1949, 1972 by Mary Boyd Higgins as Director of the
Wilhelm Reich Infant Trust
XEROX Copy from the Wilhelm Reich Museum

The Sexual Revolution
©1945, 1962 by Mary Boyd Higgins as Director
of the Wilhelm Reich Infant Trust

Zeugnisse einer Freundschaft
Der Briefwechsel zwischen Wilhelm Reich und A.S.
Neill (1936–1957)
Köln: Kiepenheuer & Witsch, 1986

Roberts, Jane

The Nature of Personal Reality
New York: Amber-Allen Publishing, 1994
First published in 1974

Die Natur der Persönlichen Realität
Ein neues Bewusstsein als Quelle der Kreativität
München: Kailash Verlag, 2007

The Nature of the Psyche
Its Human Expression
New York, Amber-Allen Publishing, 1996
First published in 1979

Die Natur der Psyche
Ihr menschlicher Ausdruck in Kreativität, Liebe, Sexualität
Genf: Ariston Verlag, 1985

Die Natur der Psyche
Ihr menschlicher Ausdruck in Kreativität, Liebe, Sexualität
München: Kailash Verlag, 2008

Rudhyar, Dane

Astrology of Personality
A Reformulation of Astrological Concepts and Ideals in Terms of Contemporary Psychology and Philosophy
New York: Aurora Press, 1990

An Astrological Triptych
Gifts of the Spirit, The Way Through, and The Illumined Road
New York: Aurora Press, 1991

Astrological Mandala
New York: Vintage Books, 1994

L'astrologie de la transformation
Paris: Rocher, 1984

Ruiz, Don Miguel

The Four Agreements
A Practical Guide to Personal Freedom
San Rafael, CA: Amber Allen Publishing, 1997

The Mastery of Love
A Practical Guide to the Art of Relationship
San Rafael, CA: Amber Allen Publishing, 1999

The Voice of Knowledge
A Practical Guide to Inner Peace
San Rafael, CA: Amber Allen Publishing, 2004

Ruperti, Alexander
Cycles of Becoming
The Planetary Pattern of Growth
New York: CRCS Publications, 1978

Sheldrake, Rupert
A New Science of Life
The Hypothesis of Morphic Resonance
Rochester: Park Street Press, 1995

Das Schöpferische Universum
Die Theorie des morphogenetischen Feldes
Neue und erweiterte Auflage
Berlin: Ullstein, 2009

Shone, Ronald
Creative Visualization
Using Imagery and Imagination for Self-Transformation
New York: Destiny Books, 1998

Smith, C. Michael
Jung and Shamanism in Dialogue
London: Trafford Publishing, 2007

Spiller, Jan
Astrology for the Soul
New York: Bantam, 1997

Thorsson, Edred
Futhark
A Handbook of Rune Magic
San Francisco: Weiser Books, 1984

Tolle, Eckhart

The Power of Now
A Guide to Spiritual Enlightenment
Novato, CA: New World Library, 2004

Jetzt! Die Kraft der Gegenwart
Ein Leitfaden zum spirituellen Erwachen
Bielefeld: Kamphausen Verlag, 2000

A New Earth
Awakening to Your Life's Purpose
New York: Michael Joseph (Penguin), 2005

Eine neue Erde
Bewusstseinssprung anstelle von Selbstzerstörung
München: Goldmann, 2005

Too, Lillian

Feng Shui
Kuala Lumpur: Konsep Books, 1994

Wild, Leon D.

The Runes Workbook
A Step-by-Step Guide to Learning the Wisdom of the Staves
San Diego: Thunder Bay Press, 2004

Wilhelm, Helmut

The Wilhelm Lectures on the Book of Changes
Princeton: Princeton University Press, 1995

Wilhelm, Richard

The I Ching or Book of Changes
With C. Baynes
3rd Edition, Bollingen Series XIX
Princeton, NJ: Princeton University Press, 1967

Williams, Strephon Kaplan

Dreams and Spiritual Growth
With Patricia H. Berne and Louis M. Savary
New York: Paulist Press, 1984

Durch Traumarbeit zum eigenen Selbst
Die Jung-Senoi Methode
Interlaken: Ansata Verlag, 1987

Dream Cards
Understand Your Dreams and Enrich Your Life
New York: Simon & Schuster (Fireside), 1991

Wing, R. L.

The I Ching Workbook
Garden City, N.Y.: Doubleday, 1984

Das Arbeitsbuch zum I Ching
Mit Chinesischen Orakel Münzen
München: Goldmann, 2004

Het I Tjing Werkboek
Baarn: Bigot & Van Rossum, 1986

Wolf, Fred Alan

Taking the Quantum Leap
The New Physics for Nonscientists
New York: Harper & Row, 1989

Der Quantensprung ist keine Hexerei
Frankfurt/M: Fischer Verlag, 1990

Parallel Universes
New York: Simon & Schuster, 1990

The Dreaming Universe
A Mind-Expanding Journey into the Realm
Where Psyche and Physics Meet
New York: Touchstone, 1995

The Eagle's Quest
A Physicist Finds the Scientific Truth At the Heart of the Shamanic World
New York: Touchstone, 1997

Die Physik der Träume
Frankfurt/M: DTV Verlag, 1997

Mind into Matter
A New Alchemy of Science and Spirit
New York: Moment Point Press, 2000

Wydra, Nancilee
Feng Shui
The Book of Cures
Lincolnwood: Contemporary Books, 1996

If there is light in the soul, there is beauty in the person. If there is beauty in the person, there is harmony in the house. If there is harmony in the house, there is order in the nation. If there is order in the nation, there will be peace on earth.
– Chinese Proverb

FROM THE SAME AUTHOR

A Bibliography

You can search publications from here:
http://ipublica.com/books/

For audio books and music, you can start here:
http://ipublica.com/audio/

All paperbacks, audio downloads, audio book compact discs, music downloads and music compact discs, as well as Kindle books, are referenced on the site.

For free podcasts search iTunes under my author name.

For quoting my publications, please use the following form:
Pierre F. Walter, [Title]: [Subtitle], Newark: Sirius-C Media Galaxy LLC, 2011

Web Presence

Pierre F. Walter on the Web

Sites

http://authoryourlife.com

http://ipublica.com

http://ipublica.net

http://ipublica.org

http://ipublica.tv

Video Channels

http://youtube.com/user/ipublica

http://youtube.com/user/authoryourlife

http://vimeo.com/pierrefwalter/channels

http://ipublica.blip.tv/

http://authoryourlife.blip.tv/

http://emosexuality.blip.tv/

http://pierrefwalter.blip.tv/

www.ingramcontent.com/pod-product-compliance
Lightning Source LLC
Chambersburg PA
CBHW021006180526
45163CB00005B/1912